# THE POETRY OF XENON

# The Poetry of Xenon

Walter the Educator

**SKB**

Silent King Books a WhichHead Imprint

Copyright © 2023 by Walter the Educator

All rights reserved. No part of this book may be reproduced in any manner whatsoever without written permission except in the case of brief quotations embodied in critical articles and reviews.

First Printing, 2023

Disclaimer
This book is a literary work; poems are not about specific persons, locations, situations, and/or circumstances unless mentioned in a historical context. This book is for entertainment and informational purposes only. The author and publisher offer this information without warranties expressed or implied. No matter the grounds, neither the author nor the publisher will be accountable for any losses, injuries, or other damages caused by the reader's use of this book. The use of this book acknowledges an understanding and acceptance of this disclaimer.

"Earning a degree in chemistry changed my life!"
- Walter the Educator

dedicated to all the chemistry lovers, like myself, across the world

# CONTENTS

Dedication . . . . . . . . . . . . . . . . . v

Why I Created This Book? . . . . . . . . . 1

**One** - Xenon, A Treasure . . . . . . . . . . . . 2

**Two** - Cosmic Symphony . . . . . . . . . . . . 4

**Three** - Xenon, The Luminary . . . . . . . . 6

**Four** - Heavenly Space . . . . . . . . . . . . . 8

**Five** - Enlivens Our Souls . . . . . . . . . . . 10

**Six** - Xenon, The Radiant . . . . . . . . . . . 12

**Seven** - Curiosity Survives . . . . . . . . . . 14

**Eight** - Guiding Force . . . . . . . . . . . . . 16

**Nine** - Truly Own . . . . . . . . . . . . . . . . 18

**Ten** - More Than Gold . . . . . . . . . . . . . 20

**Eleven** - Day By Day . . . . . . . . . . . . . 22

**Twelve** - Embrace The Unknown . . . . . . 24

| | | |
|---|---|---|
| **Thirteen** - Unique And Divine | . . . . . . . . | 26 |
| **Fourteen** - Behold | . . . . . . . . | 28 |
| **Fifteen** - Xenon, The Noble Gas | . . . . . . | 30 |
| **Sixteen** - Unlock The Secrets | . . . . . . . . | 32 |
| **Seventeen** - A Poet's Dream | . . . . . . . . | 34 |
| **Eighteen** - Source Of Curiosity | . . . . . . . . | 36 |
| **Nineteen** - For Xenon | . . . . . . . . | 38 |
| **Twenty** - Testament | . . . . . . . . | 40 |
| **Twenty-One** - Captivating | . . . . . . . . | 42 |
| **Twenty-Two** - Silent Warrior | . . . . . . . . | 44 |
| **Twenty-Three** - Your Brilliance | . . . . . . . . | 46 |
| **Twenty-Four** - We Adore | . . . . . . . . | 48 |
| **Twenty-Five** - Cherish You | . . . . . . . . | 50 |
| **Twenty-Six** - Glow | . . . . . . . . | 52 |
| **Twenty-Seven** - Dear Xenon | . . . . . . . . | 54 |
| **Twenty-Eight** - Knowledge | . . . . . . . . | 56 |
| **Twenty-Nine** - Touches Our Lives | . . . . . . . . | 58 |
| **Thirty** - It Shines | . . . . . . . . | 60 |
| **Thirty-One** - Abides | . . . . . . . . | 62 |
| **Thirty-Two** - Source Of Inspiration | . . . . . | 64 |

**Thirty-Three** - Illuminates Our World . . . . . 66

**Thirty-Four** - Oh Xenon . . . . . . . . . . . . 68

About The Author . . . . . . . . . . . . . . 70

# WHY I CREATED THIS BOOK?

Creating a poetry book about the chemical element Xenon was a fascinating and unique endeavor. Xenon, with its rare and mysterious properties, offers a rich source of inspiration. By exploring the various aspects of Xenon, such as its physical characteristics, its role in the universe, and its potential applications, a poet can create a collection that delves into themes of science, nature, human existence, and even metaphysics. The juxtaposition of scientific concepts with the beauty of poetry can lead to a thought-provoking and captivating exploration of the element Xenon.

# ONE

# XENON, A TREASURE

In the cosmos' dance of elements rare,
A noble gas, Xenon, floats through the air.
With atomic number fifty-four it stands,
A luminescent gem in nature's hands.

Its name derived from the Greek, xenos,
A stranger to the world, yet it bestows
A vivid light in hues that mesmerize,
A glow that enchants and mystifies.

In gas form, it remains colorless, serene,
But under pressure, it shines with a sheen.
A beacon of brilliance, it imparts its glow,
Illuminating the dark with a cosmic show.

Within the Earth, it hides in the deep,
Where ancient whispers of time it keeps.

Unreactive, it stands aloof and still,
A symbol of calm, a tranquil thrill.
    Yet in the realm of science, it finds a role,
Aiding in lasers and lighting the soul.
Its radiance guides us through the night,
A testament to nature's wondrous might.
    Xenon, a treasure from the celestial sphere,
A silent luminary, gentle and clear.
May we forever cherish its ethereal grace,
And bask in the glow of its cosmic embrace.

# TWO

# COSMIC SYMPHONY

In a realm beyond, where stars align,
There dwells a noble gas, so divine.
Xenon, its name, from Greek it derives,
A stranger, enchanting, as it thrives.

A silent luminary, it gracefully shines,
Embracing the cosmos with ethereal lines.
A treasure from celestial sphere,
Xenon's presence, so rare and dear.

Under pressure, it reveals its might,
A beacon of brilliance, a guiding light.
In lasers it dances, with precision and grace,
Illuminating the path, leaving no trace.

Oh Xenon, bearer of scientific advance,
In lighting the way, you enhance.

From lamps that flicker, to screens that glow,
You bring radiance, wherever you go.
    Unseen by many, yet so profound,
Xenon's essence, a celestial sound.
Embrace its beauty, like a starlit night,
Let its enchantment fill you with delight.
    Oh Xenon, you're truly a gem,
A noble gas, a captivating emblem.
With elegance and grace, you ignite,
A cosmic symphony, a celestial light.

# THREE

# XENON, THE LUMINARY

In the depths of the cosmos, a gem does reside,
A luminescent wonder, with secrets inside.
Xenon, the element, enchanting and rare,
A celestial jewel, beyond compare.

From distant stars, it traveled far and wide,
To grace our world, with its ethereal pride.
It dances in darkness, its glow ever bright,
A beacon of radiance, in the blackest of night.

Oh Xenon, you captivate our curious minds,
With your atomic beauty, that forever binds.
In noble gases, you hold your noble reign,
A symbol of brilliance, in the scientific domain.

In tubes and bulbs, you bring forth your light,
Illuminating the shadows, banishing the night.

Your spectral lines, a kaleidoscope of hues,
A symphony of colors, that forever amuse.

In lasers and flashlights, you lend us your power,
A gift from the heavens, in our darkest hour.
You guide our research, our breakthroughs, our quest,
Unveiling the secrets, we yearn to digest.

Oh Xenon, celestial gem, forever sublime,
Your luminous presence, a gift, so divine.
In laboratories and beyond, your magic prevails,
A testament to science, where wonder never fails.

So let us rejoice, in Xenon's embrace,
A celestial treasure, adorning time and space.
With every flicker, every shimmer, every glow,
Xenon, the luminary, forever aglow.

# FOUR

## HEAVENLY SPACE

In the realm of elements, a jewel does gleam,
A noble gas, Xenon, its presence supreme.
With atomic number fifty-four, it lies,
A beacon of light, amidst the starry skies.
    Mysterious and rare, it does abide,
In nature's embrace, where secrets hide.
A luminescent glow, it does possess,
A shimmering brilliance, none can suppress.
    In science's realm, it plays a vital role,
A noble gas, with stories yet untold.
From lighting the way in bright xenon lamps,
To aiding in lasers, with radiant stamps.
    Xenon, the celestial traveler's guide,
Guiding explorers on their cosmic ride.

Through distant galaxies, it does dance,
Enlightening the cosmos with its chance.

    Beyond the boundaries of Earth's embrace,
Xenon journeys into the vastness of space.
Its rarity cherished, its flare adored,
A testament to science's rewarding reward.

    Oh, Xenon, the enigmatic and pure,
In darkness, you shine, a luminary allure.
A key to the mysteries our world conceals,
Unraveling secrets, your brilliance reveals.

    So let us marvel at this noble gas,
And toast to the wonders it brings en masse.
Xenon, the element of light and grace,
Forever glowing in its heavenly space.

# FIVE

# ENLIVENS OUR SOULS

In the cosmic realm, a luminescent celestial ember,
A radiant gem, the essence of Xenon I remember.
With noble grace, it dances in the depths of space,
A starlit haze, its ethereal glow I embrace.

Oh, Xenon, your brilliance ignites the darkest night,
A beacon of light, guiding seekers towards insight.
In laboratories, you reveal secrets untold,
Unraveling mysteries, a scientist's stronghold.

From the depths of the Earth, you rise with gentle might,
A treasure concealed, a rare and precious sight.
In noble gases, you hold a steady place,
An element of grace, embracing time and space.

With an azure glow, you captivate the eye,
A mystic hue that paints the evening sky.

Your electrons dance, in a delicate symphony,
A cosmic melody, in perfect harmony.
    Oh, Xenon, you shine with ethereal allure,
A celestial treasure, forever I'll adore.
In science and nature, your presence unfolds,
A story untold, that forever enlivens our souls.

# SIX

# XENON, THE RADIANT

In the realm of elements, a beauty does reside,
A noble gas of wonder, with secrets held inside.
Xenon, the ethereal, with brilliance all its own,
A shimmering luminary, in a world where shadows roam.

In laboratories it dances, with atoms it does play,
A puzzle for the scientists, in its mysterious array.
Its electrons, they are stable, in a valence so inert,
Yet it holds a power untamed, a light that can't be hurt.

From lighting to lasers, its presence does illuminate,
A guiding force in darkness, a beacon to captivate.
Its glow, so pure and vibrant, a celestial display,
Xenon's ethereal essence, forever lights the way.

In the depths of exploration, it guides us through

the night,
Unraveling the mysteries, with its scientific might.
From distant distant stars to the depths of the unknown,
Xenon opens doors to realms, where secrets are sown.

 Oh, Xenon, the enigmatic, with brilliance in your core,
You captivate our senses, like none have done before.
A treasure among elements, a precious gem to find,
Xenon, the radiant, forever etched in our mind.

# SEVEN

# CURIOSITY SURVIVES

In the realm of elements, a gem so rare,
Xenon, the beacon of brilliance, fair.
Its noble nature, calm and serene,
A symbol of tranquility, seldom seen.

Amidst the periodic table's grand design,
Xenon, a star, with elegance, does shine.
A cosmic symphony, it softly plays,
Unveiling secrets in illuminating ways.

With electrons dancing in its outer shell,
Xenon, the celestial traveler, can tell
Tales of distant worlds, unknown and vast,
Guiding us through darkness, holding steadfast.

Its radiance captured in the depths of light,
Xenon, a celestial gem shining bright.
In lasers and lighting, it finds its place,
Casting its glow with grace and endless grace.

Oh, Xenon, your beauty, a wondrous sight,
A noble gas, a marvel of the night.
Through your ethereal glow, mysteries unwind,
Revealing the wonders of the universe, aligned.
So let us embrace this element divine,
Xenon, the guide, leading us to find
The realms of the unknown, where knowledge thrives,
With your brilliance, our curiosity survives.

# EIGHT

# GUIDING FORCE

In the depths of darkness, where shadows reside,
There lies a luminary, a celestial guide.
Xenon, the element of mystery and might,
Unraveling secrets, revealing the light.

A noble gas, rare and ethereal,
A beacon of brilliance, both bold and surreal.
In the stillness of the night, it softly glows,
A shimmering star, where enchantment flows.

Within the depths of a laboratory's embrace,
Xenon dances with electrons, creating a trace.
In lighting and lasers, its presence is found,
A spectacle of radiance, astoundingly profound.

From distant galaxies to the farthest frontiers,
Xenon traverses the cosmos, conquering fears.
As it journeys through space, it captivates the eye,
A celestial traveler, painting the sky.

In scientific exploration, it plays a crucial role,
Unveiling the universe, expanding the soul.
With every discovery, its power is revealed,
A catalyst for knowledge, a treasure unsealed.
   Xenon, the element of wonder and surprise,
A symphony of atoms, a celestial prize.
Forever it shines, in darkness and in day,
A guiding force, lighting the way.

# NINE

# TRULY OWN

    Xenon, a kaleidoscope of colors rare,
A gift from heavens, beyond compare.
In noble gases, it holds its place,
Guiding research, unveiling grace.
    From deep within the earth it's found,
A treasure hidden, under the ground.
With atomic number fifty-four,
Xenon's mysteries we'll explore.
    A celestial traveler, it takes its flight,
Enlightening the cosmos with its light.
Through the vast expanse of space it roams,
Revealing secrets, it calls its own.
    In laboratories, its power revealed,
A guiding force, it won't be concealed.
With luminescent glow, it shines bright,
Unraveling the mysteries of the night.

Among the noble gases, it stands tall,
A steady presence amidst them all.
With its noble heart, it leads the way,
Unveiling truths, day by day.

Xenon, a beacon of light so pure,
Illuminating darkness, that's for sure.
Opening doors to unknown realms,
Where wonders lie, and wisdom overwhelms.

A celestial traveler, captivating with its radiance,
Guiding us through the universe, with elegance.
Revealing the wonders of the unknown,
Xenon, a treasure we truly own.

# TEN

# MORE THAN GOLD

In the realm of scientific exploration,
Where mysteries unravel in contemplation,
There lies a noble element, Xenon by name,
A beacon of light, a celestial flame.

With its atomic number, fifty-four,
Xenon illuminates the deep cosmic core,
A noble gas, noble in its essence,
Shining brightly, defying pretense.

Through the depths of darkness, it does glide,
In noble silence, it does confide,
Its radiance captivating, a cosmic dance,
A luminescent glow, a celestial trance.

In lighting and lasers, it finds its place,
A luminary presence, full of grace,

A guiding force, a beacon of might,
Xenon's brilliance, a celestial light.
   From the depths of the universe, it travels afar,
Guiding explorers, like a guiding star,
Unraveling secrets, unlocking the unknown,
Xenon's brilliance, forever shown.
   Oh, Xenon, element of wonder and awe,
A celestial traveler, beyond natural law,
With every atom, a story untold,
Xenon, a treasure, worth more than gold.

# ELEVEN

# DAY BY DAY

In the realm of the unknown, a light does gleam,
A noble gas, Xenon, the stuff of dreams.
Unseen and silent, it guides the way,
Unraveling mysteries, night and day.

In the heart of lighting, its brilliance shines,
A beacon of radiance, so divine.
In lasers it dances, with grace untold,
Revealing wonders, that were once untold.

Oh Xenon, ethereal and pure,
A captivating allure, forever secure.
A guiding force, through the darkest haze,
Leading us towards the celestial maze.

Through scientific exploration, we soar,
With Xenon as our compass, we implore,

To discover the secrets of the universe vast,
With each revelation, a spell is cast.

    Oh Xenon, you are a marvel to behold,
A testament to the wonders yet untold.
In your presence, we find solace and light,
Guiding us through the mysteries of the night.

    So let us celebrate this noble gas,
And the cosmic wonders that it does amass.
For in Xenon's essence, we find our way,
To unravel the universe, day by day.

# TWELVE

# EMBRACE THE UNKNOWN

In the realm of luminescence, Xenon reigns supreme,
A radiant noble gas, a poet's cherished dream.
Its atomic number fifty-four, a symbol of the stars,
Xenon's ethereal glow guides us past the cosmic bars.

With a flicker of brilliance, it dances in the night,
A celestial beacon, a source of pure light.
From neon signs to lasers, its radiance takes form,
Illuminating our world, defying the norm.

In the depths of laboratories, scientists delve,
Exploring Xenon's secrets, the stories it can tell.
Its properties unique, its mysteries untold,
A key to understanding the universe's mold.

Xenon, the catalyst of scientific revelation,
Unraveling the cosmos with unwavering dedication.

From noble experiments to groundbreaking feats,
It unlocks the doors where knowledge retreats.

Oh Xenon, celestial wanderer, guide us on our quest,
Through the vast expanse of space, we are truly blessed.
With your shimmering brilliance, you show us the way,
To uncover the wonders that in darkness lay.

So let us celebrate Xenon, a marvel of creation,
A symbol of curiosity and scientific exploration.
Its captivating beauty, its radiant embrace,
Inspires us to seek knowledge and embrace the unknown space.

# THIRTEEN

# UNIQUE AND DIVINE

In the celestial realm, a traveler resides,
A beacon of light, where darkness subsides.
Xenon, the element, so rare and sublime,
A guiding force through the fabric of time.

With electrons aglow, in the depths it glimmers,
Unraveling mysteries, like celestial whispers.
A catalyst for knowledge, it sparks curiosity,
Expanding horizons, with scientific ferocity.

Oh Xenon, your luminescent hue,
A symphony of colors, forever anew.
From neon to blue, a kaleidoscope dance,
Enchanting our senses, in a captivating trance.

Through lenses and lasers, your power we harness,
Unveiling the wonders, in Nature's vastness.

In noble gases, you reign supreme,
A treasure more precious than gold, it seems.
   In laboratories, you pave the way,
For discoveries that shape the world today.
From medical marvels to lighting the stage,
Xenon, your impact, knows no gauge.
   So let us celebrate this remarkable gas,
A celestial traveler, that inspires en masse.
In the realm of science, you'll forever shine,
Xenon, the element, so unique and divine.

# FOURTEEN

## BEHOLD

In the realm of luminescent glow,
A noble gas, Xenon does bestow.
With atomic number fifty-four,
It guides us through the universe's door.

   A beacon of light in the cosmic sea,
Xenon shines with ethereal glee.
Its electrons dance, oh, how they gleam,
Illuminating the spaces in between.

   Through scientific exploration, we find,
Xenon's secrets, so beautifully aligned.
Unraveling mysteries, one by one,
In the realm of atoms, it's second to none.

   A treasure among the gases, so rare,
Xenon, with a presence beyond compare.

Noble in nature, noble in name,
It holds the key to a scientific flame.
    Radiant and captivating, it lures,
With its luminescent allure.
From the depths of the Earth to the stars above,
Xenon reveals the wonders we seek to prove.
    Oh, Xenon, a catalyst of revelation,
Guiding us through the ages of exploration.
In laboratories, it sparks new creation,
Unleashing the power of scientific elation.
    So let us celebrate this element divine,
Xenon, a treasure so sublime.
With its radiant glow and mysteries untold,
It's a beacon of knowledge we forever behold.

# FIFTEEN

# XENON, THE NOBLE GAS

In the realm of noble gases, a luminescent star,
Xenon, the element, with brilliance from afar.
A shimmering allure, captivating and bright,
In the depths of exploration, it guides our sight.

With electrons tightly bound, it stands alone,
Inert and noble, its character proudly shown.
In the vast expanse of the cosmos, it prevails,
Unraveling mysteries, its luminescence unveils.

Xenon, a beacon of discovery and awe,
A testament to the wonders that science can draw.
Its noble presence, a guiding light through the night,
Igniting curiosity, filling hearts with delight.

From the depths of Earth's atmosphere it is found,
A treasure waiting to be explored, to astound.

With its spectral lines, it reveals the unseen,
Unlocking the secrets of the universe, serene.

Oh, Xenon, you dazzle with your atomic glow,
A captivating beauty, a celestial show.
Through your noble essence, we find our way,
Inspiring us to seek knowledge, day by day.

Let us celebrate Xenon, our scientific guide,
A luminescent element, forever by our side.
In the realm of discovery, it will forever shine,
Xenon, the noble gas, a treasure so divine.

# SIXTEEN

# UNLOCK THE SECRETS

Xenon, the guiding force of light,
A celestial traveler in the darkest night.
With noble grace, you float through the air,
A catalyst for scientific minds to stare.

In your shimmering brilliance, you glow,
A beacon of knowledge, for all to know.
Your atomic secrets, hidden from view,
Revealed by the curious, the chosen few.

Oh Xenon, your beauty captivates the gaze,
As scientists explore, in endless ways.
Through prisms of wonder, your colors dance,
A symphony of hues, a mesmerizing trance.

In the realm of medicine, you find your place,
Unleashing marvels with your healing grace.
From anesthesia's realm, you gently descend,
Easing the pain, a doctor's trusted friend.

Rare among gases, you hold your ground,
An enigmatic element, so rarely found.
Yet, in your scarcity, lies a cosmic truth,
That the universe holds secrets, yet to soothe.
Xenon, noble presence, ignite our curiosity,
Unveil the universe's enigmatic tapestry.
In your essence, we find the key,
To unlock the secrets, hidden for eternity.

# SEVENTEEN

# A POET'S DREAM

In the depths of the cosmos, a noble light does shine,
Xenon, the element rare, a treasure so divine.
With a luminescent glow, it guides us through the night,
Unveiling the mysteries, filling our hearts with delight.

A whisper in the darkness, a beacon in the sky,
Xenon's ethereal beauty, no mortal can deny.
In noble gas it dwells, so tranquil and serene,
A dance of electrons, a spectacle unforeseen.

Through the lenses of science, it reveals its hidden might,
Aiding in exploration, pushing boundaries to new heights.

In lasers it finds purpose, cutting through the haze,
Unleashing its power, in a mesmerizing blaze.

In the realm of medicine, it lends a helping hand,
A diagnostic tool, a healer in demand.
From MRI machines, to anesthesia's embrace,
Xenon's healing touch, brings solace to the human race.

And as the world slumbers, beneath the starlit sky,
Xenon lights our way, with a gentle, watchful eye.
In bulbs it finds its calling, illuminating our nights,
A soft and steady glow, bringing warmth and delight.

So let us celebrate, this element so rare,
Xenon, the guiding light, beyond compare.
In the realm of chemistry, it reigns supreme,
A testament to nature's brilliance, a poet's dream.

# EIGHTEEN

# SOURCE OF CURIOSITY

In Xenon's realm, a dance of light unfolds,
A kaleidoscope of colors, stories untold.
A noble gas, mysterious and rare,
Xenon, a wonder beyond compare.

In laboratories, its secrets are unveiled,
Its properties, by scientists, hailed.
From glowing tubes to shimmering lamps,
Xenon's radiance, like celestial camps.

In medicine's embrace, it finds its place,
A healer, a guide, with gentle grace.
Diagnostic tools, with Xenon's aid,
Reveal the body's mysteries, unswayed.

Anesthesia's ally, a comforting hand,
Xenon's touch, a tranquil land.

In surgical realms, where pain resides,
Xenon's solace, where darkness subsides.

And in the darkest corners of the night,
Xenon's glow, a beacon of light.
Illuminating cities, streets, and homes,
Xenon's warmth, wherever it roams.

Oh Xenon, enigmatic and profound,
You captivate, wherever you're found.
A silent messenger, unlocking the skies,
Revealing the universe, before our eyes.

So let us celebrate this element divine,
Xenon, a treasure, forever enshrined.
In science, medicine, and the world we see,
Xenon, a source of curiosity.

# NINETEEN

# FOR XENON

In the depths of the universe, a treasure lies,
A rare and noble element, Xenon, it flies.
A beacon of light, it illuminates the night,
Unveiling the secrets, hiding out of sight.

With noble gases, it shares its noble creed,
A symbol of enlightenment, a guide we need.
Its atomic number, a testament to its might,
Xenon, the element that radiates pure light.

In medicine's realm, it finds its sanctuary,
A healer, a savior, with powers extraordinary.
From the operating room to the neonatal ward,
Xenon's touch brings solace, whispers hope restored.

A protector of minds, it guards memories dear,
Fighting against darkness, erasing all fear.
In the depths of anesthesia, it lends its hand,
Guiding the unconscious, through an ethereal land.

A puzzle to scientists, it captivates their gaze,
Unlocking the mysteries, in a thousand ways.
With its spectral lines, it paints the cosmos' tale,
A cosmic fingerprint, on a celestial trail.

Oh, Xenon, rare gem, in the realm of the small,
You hold the key, to understand it all.
In laboratories, your secrets we uncover,
In your noble essence, we find wonder.

So let us celebrate, this element divine,
For Xenon, the guiding light, forever shines.

# TWENTY

# TESTAMENT

In the realm of elements, rare and bright,
There lies a gem, a radiant light.
Xenon, noble gas, so serene,
A marvel of nature, a sight to be seen.
    In lasers it dances, with precision and grace,
Its photons entwined, in a shimmering embrace.
From green to blue, its hues doth change,
In pulses of energy, it rearranges.
    In medicine's realm, it finds its place,
Anesthesia's partner, with gentle embrace.
Calming the nerves, easing the pain,
Xenon, the healer, a solace regained.
    And when the night falls, and darkness descends,
Xenon illuminates, with a glow that transcends.

Streetlights and headlights, shining so bright,
Guiding our way, through the darkest of night.

But beyond its uses, practical and grand,
Xenon protects memories, with a steadfast hand.
In cameras and flashes, it captures the past,
Preserving our stories, memories that last.

Oh Xenon, noble gas, so rare and unique,
A shining beacon, a scientific mystique.
In laboratories and experiments, you shine so bright,
A guiding light in the realm of science's might.

So let us celebrate this element divine,
Xenon, the gem, that forever will shine.
In the periodic table, a treasure to behold,
A testament to nature's wonders untold.

# TWENTY-ONE

# CAPTIVATING

In the realm of elements, a noble gas divine,
Xenon, mysterious, in its own cosmic design.
Bathed in ethereal glow, it illuminates the night,
A beacon of radiance, a celestial light.

Xenon, the enigma, holds secrets untold,
Unleashing its power, unlocking the manifold.
In lasers, it dances, a graceful display,
Harnessing its energy, in a vivid array.

In the realms of medicine, it lends its gentle touch,
Anesthesia's companion, comforting much.
Soothing the soul, calming the mind,
Xenon's embrace, a solace we find.

In lighting's embrace, it casts a gentle hue,
A soft luminescence, a warm welcome to view.
In theaters and homes, its glow does enhance,
Creating an ambiance, a captivating trance.

Yet beyond these realms, it ventures to explore,
Unraveling the universe, its secrets to restore.
In scientific research, it aids the quest,
Xenon's wisdom, guiding us to our best.

Oh Xenon, the element, so versatile and grand,
With every facet, a marvel, in its own command.
From lighting to lasers, medicine to the sky,
Xenon, forever captivating, as time goes by.

# TWENTY-TWO

# SILENT WARRIOR

In the realm of science, a noble gas resides,
With an ethereal glow, where wonder coincides,
Xenon, the element, mysterious and rare,
A captivating essence that permeates the air.
   In the realm of medicine, its power takes flight,
As an agent of anesthesia, casting darkness to light,
Through its gentle touch, consciousness is suppressed,
And pain's cruel grip is gently laid to rest.
   In the realm of the night, it dances and glows,
A beacon of luminescence, as the darkness bestows,
The neon signs flicker with a vibrant hue,
Xenon's brilliance illuminates the world anew.
   In the realm of research, it aids the quest for truth,
As a tool for spectroscopy, revealing nature's sleuth,

Its spectral lines unveil the secrets of the stars,
Unraveling the mysteries that lie afar.

Xenon, the enigma, with its atomic might,
Preserving memories, capturing the past in a light,
Its noble presence, a testament to its worth,
Shaping the world in ways that transcend the Earth.

So let us marvel at this wondrous element's grace,
In lighting our path and illuminating space,
Xenon, a versatile gem, with talents untold,
A silent warrior, shaping the world, bold.

# TWENTY-THREE

# YOUR BRILLIANCE

In the realm of science, a hidden gem,
A noble gas, a whisper in the wind.
Xenon, the element, so rare and pure,
With mysteries and wonders yet to endure.

In the depths of darkness, it finds its place,
Anesthesia's ally, in tranquil grace.
It soothes the souls, it eases the pain,
As consciousness drifts, like a gentle rain.

But beyond the realm of medicine's might,
Xenon illuminates the starry night.
In electric bulbs, its glow so bright,
It banishes shadows, brings forth the light.

And in the realm of scientific quest,
Xenon, the element, outshines the rest.

With its spectral dance, it reveals the stars,
Unveiling secrets, the universe unbars.
   But let us not forget its vibrant role,
In neon signs, it takes a vital stroll.
A brilliant display, a mesmerizing sight,
Xenon's brilliance shapes the world, so bright.
   So let us celebrate this noble gas,
Its properties unique, its power vast.
Xenon, the element, we raise our glass,
To your brilliance and wonder, we amass.

# TWENTY-FOUR

# WE ADORE

Xenon, the illuminating light,
A noble gas, shining so bright.
In laboratories, you're put to the test,
Unveiling secrets, where truth is blessed.

In the dark of night, you come alive,
A beacon of hope, you strive to thrive.
Your glow, serene and mesmerizing,
Guides lost souls, gently enticing.

In medicine, you play a vital role,
Preserving memories, healing the soul.
Through imaging techniques, you unveil,
The mysteries of life, like a fairytale.

In the vast universe, you leave your mark,
Unraveling the secrets hidden in the dark.
From distant stars to cosmic rays,
You illuminate the path, in wondrous ways.

Neon signs, adorned with your grace,
Creating art in every bustling place.
Your vibrant colors, a breathtaking sight,
Captivating hearts, shining so bright.

Xenon, oh Xenon, you light the way,
Banishing shadows, night and day.
A catalyst for science, art, and more,
Your brilliance, forever we adore.

# TWENTY-FIVE

# CHERISH YOU

In the realm of noble gases, a marvel stands alone,
Xenon, the radiant element, with a brilliance all its own.
With noble grace and luminescence bright,
It bathes the world in a captivating light.

In the streets, it dances within the night,
Streetlights flicker, casting shadows out of sight.
Guiding our steps, it illuminates the way,
A beacon of safety until the break of day.

In theater halls, it takes center stage,
Casting a glow, enchanting the audience with its sage.
Neon signs come alive with vibrant hues,
Drawing us closer, like a moth to the flame it ensues.

In science's realm, it holds secrets untold,
Spectroscopy reveals its stories, so bold.

Through its spectral lines, knowledge is found,
Unraveling the universe, with each atom's sound.

Yet, Xenon's allure extends beyond the realm of science,
For its powers embrace the essence of human reliance.
In memory preservation, it plays a vital role,
Capturing moments, preserving them, making us whole.

And in medical imaging, it heals the soul,
A luminescent journey, making us whole.
With Xenon's touch, the body is revealed,
Diagnosis and healing, its power concealed.

Oh, Xenon, the wondrous element divine,
In your radiance, the world does shine.
A celestial gift, both rare and true,
In our hearts, forever, we shall cherish you.

# TWENTY-SIX

## GLOW

In the depths of the periodic table's scroll,
Lies a noble gas, mysterious and bold.
Xenon, the element, so versatile and rare,
A gem of the elements, beyond compare.

In the realm of memory, it finds its grace,
Preserving moments, time cannot erase.
A whisper of the past, locked within its core,
Xenon's essence, forever to explore.

In the realm of medicine, it takes its flight,
Aiding doctors, illuminating their sight.
Magnetic resonance imaging, its divine art,
Unraveling secrets, healing every heart.

In the realm of science, it dances with delight,
Its spectral lines revealing the universe's might.

Through prisms of vibrant hues, it unveils,
The cosmic wonders, where our curiosity sails.

In the realm of art, it paints with ethereal glow,
A luminescent muse, a radiant show.
A flicker of brilliance, a vibrant hue,
Xenon's light, igniting creativity anew.

Oh, Xenon, you shine with such grace,
In every field, a luminary's embrace.
From science to art, and medicine's call,
You illuminate our world, standing tall.

So let us celebrate your brilliance, dear Xenon,
A beacon of light, in a world unknown.
A symbol of wonders, both seen and unseen,
In your radiant glow, the universe we gleam.

# TWENTY-SEVEN

# DEAR XENON

In the realm of science, where wonders unfold,
There lies a noble element, a tale yet untold.
Xenon, the treasure, so rare and so pure,
A luminescent gem, with mysteries to endure.

In laboratories it dances, with atoms in stride,
Aiding the curious minds, in their quest to confide.
Its brilliance, a beacon, in the realm of the small,
Unveiling the secrets, that science enthralls.

From lighting the night with a soft neon glow,
To lasers that guide us, where we dare not go.
Xenon, the artist, with colors so bright,
Painting the canvas of the darkness of night.

But its prowess extends, beyond scientific lore,
Into realms of healing and memories it store.
For in medicine's realm, it finds its own grace,
Preserving our memories, time cannot erase.

And in the world of art, where expression takes flight,
Xenon, the muse, brings visions to light.
With noble gas tubes, it illuminates the signs,
Guiding lost souls, with a glow that defines.

Oh, Xenon, the enigma, in your noble form,
You captivate our hearts, in a brilliance so warm.
From the depths of the stars, to the touch of our souls,
Your presence, dear Xenon, forever consoles.

# TWENTY-EIGHT

# KNOWLEDGE

Xenon, noble gas of scientific might,
In labs and research, you shine so bright.
With atomic number fifty-four you stand,
A symbol of knowledge, a guiding hand.

In lighting's realm, your glow we adore,
Neon signs and bulbs, you enchant and restore.
A luminescent dance, a mesmerizing sight,
Illuminating our world with purest light.

But beyond the glow, a deeper tale unfolds,
In medicine's realm, your worth untold.
Radiology's ally, you aid in diagnosis,
Revealing secrets hidden within us.

In memory's realm, you hold a key,
Preserving thoughts for eternity.
Anesthetic of choice, you calm and sedate,
Easing pain and anxiety, our fears abate.

Oh Xenon, element of mystery and grace,
In science, art, and healing, you find your place.
From spectral lines to creative inspiration,
You touch our lives with your wondrous sensation.
   So let us marvel at your noble might,
Xenon, element of endless light.
In your brilliance, we find solace and hope,
A beacon of knowledge, a radiant scope.

# TWENTY-NINE

# TOUCHES OUR LIVES

In a realm where darkness reigns,
A shimmering light, Xenon, sustains.
An element rare, noble and bright,
Paints the night sky with celestial might.

From the depths of Earth, it takes its flight,
A beacon of wonder, a cosmic delight.
With a name that echoes ancient lore,
Xenon, the explorer, forevermore.

In tubes of glass, it dances and glows,
Neon signs whisper secrets it knows.
A rainbow of colors, vibrant and bold,
Xenon's presence, a story untold.

But beyond the signs and the neon gleam,
Xenon's magic is not what it seems.

In the realms of science, it finds its place,
Unveiling mysteries, with every trace.

In medicine's realm, it lends a hand,
Imaging techniques, a diagnostic brand.
Through veins and arteries, it softly flows,
Preserving memories, where healing grows.

In art's embrace, it ignites the soul,
Inspiring visions, making them whole.
A muse for creators, a palette of dreams,
Xenon's essence, the artist redeems.

In healing's touch, it brings relief,
Calming the mind, soothing the grief.
A therapeutic ally, gentle and kind,
Xenon's presence, a solace to find.

So let us celebrate this element rare,
Xenon, the wonder, beyond compare.
From science to art, from healing to light,
It touches our lives, with radiance bright.

# THIRTY

# IT SHINES

In the depths of darkness, a noble gas gleams,
Xenon, the enigma, in ethereal dreams.
A whisper of secrets it softly reveals,
A luminary of wonder, its aura conceals.
  With spectral glow, it dances in the night,
A beacon of brilliance, a celestial light.
Mysterious and rare, it captures the eye,
Xenon, the mystic, soaring so high.
  In medicine's realm, it lends a healing hand,
An ally in diagnosis, a diagnostic brand.
Preserving our memories, it gently bestows,
A guardian of thoughts, where remembrance flows.
  In art's vibrant canvas, it sparks inspiration,
Igniting the imagination, a creative sensation.

Illuminating signs with its radiant shine,
Xenon, the muse, where creativity aligns.
  With tranquil embrace, it calms and sedates,
A refuge of solace, where peace permeates.
Bringing relief to the weary and pained,
Xenon, the healer, whispers serenity attained.
  In imaging techniques, it paints vivid scenes,
Revealing the unseen, as reality convenes.
Therapeutic in nature, it holds a power,
To mend and restore, in its luminous shower.
  Xenon, the guardian, a symbol of light,
Guiding us forward, through the darkest of night.
A catalyst of knowledge, a radiant guide,
Forever it shines, in our world so wide.

# THIRTY-ONE

# ABIDES

In ethereal realms where dreams reside,
A shimmering light, by science, is eyed,
Xenon, the element, rare and profound,
A beacon of brilliance, the universe found.

In labs of discovery, it dances and gleams,
A noble gas, with ethereal beams,
Its electrons, they dance, in orbits so wide,
With a luminescent glow, that cannot hide.

In the depths of the night, it paints the sky,
Illuminating stars, as they pass by,
A cosmic kaleidoscope, an artist's delight,
Xenon's essence, a canvas so bright.

In the realm of medicine, it holds a key,
Anesthesia's ally, in surgery's decree,

A gentle sedation, it brings to the soul,
Healing the wounded, making them whole.

    But beyond the realms of science and art,
Xenon touches lives, it mends the heart,
A symbol of hope, in a world full of strife,
Guiding us towards a more enlightened life.

    So let us celebrate, this noble gas divine,
Xenon, the element, so rare and fine,
An illuminating force, in our world it resides,
A testament to beauty, where knowledge abides.

# THIRTY-TWO

## SOURCE OF INSPIRATION

In the realm of medicine, a healer's delight,
Xenon emerges, a radiant light.
A noble gas, mysterious and rare,
With healing properties beyond compare.

Through veins it flows, a curative grace,
Easing pain, bringing comfort, leaving no trace.
A gentle touch, a soothing embrace,
Xenon's magic, a gift from space.

In the artist's realm, a stroke of the brush,
Xenon's brilliance, a palette to hush.
A vibrant hue, an ethereal glow,
Inspiring creations that forever flow.

In the depths of the mind, a tranquil guide,
Xenon whispers, fears and worries subside.

A calm sedation, a peaceful embrace,
A respite from chaos, a tranquil space.

   In the realm of science, a guiding light,
Xenon reveals secrets, shining so bright.
With each radiology scan, its brilliance unfurls,
Mapping the body, painting a world.

   In the realm of memories, a treasure to keep,
Xenon preserves moments, memories deep.
A guardian of time, a preserver of years,
Ensuring the past forever appears.

   In the realm of surgery, a gentle touch,
Xenon's anesthesia, a blessing, not much.
A serene slumber, a painless reprieve,
Guiding the hand to bring healing, believe.

   So let us marvel at Xenon's might,
A symbol of hope, a beam of light.
An element rare, yet found within,
A source of inspiration, a pathway to begin.

# THIRTY-THREE

# ILLUMINATES OUR WORLD

In realms of radiance, Xenon gleams,
A luminary born of noble dreams.
With gentle touch, it paints the night,
A beacon in darkness, shining bright.

In chambers of science, it finds its place,
Revealing secrets, unveiling grace.
A tool for healing, a friend to the ill,
Xenon's powers, a gift to fulfill.

In radiology's realm, it takes its stand,
Guiding the way with an expert hand.
Through shadows it travels, with steadfast will,
Unveiling the truth, a vision to instill.

In memories' realm, it finds its space,
Preserving moments, a treasured embrace.

Whispering softly, it captures the past,
A solace for hearts that want it to last.
    In anesthesia's realm, it casts its spell,
Easing the pain, a tranquil farewell.
Soothing the soul, in slumber it holds,
A passage to dreams, where peace unfolds.
    Xenon, a rarity, both ethereal and true,
A friend to the lost, a guide to the few.
In art and in science, it plays its part,
A masterpiece of nature, a work of fine art.
    So let us embrace this luminescent gem,
With gratitude and awe, we honor them.
For Xenon, the element, rare and divine,
Illuminates our world, makes it truly shine.

# THIRTY-FOUR

## OH XENON

In the realm of the noble gases, there lies
A luminescent jewel, radiant and wise.
Xenon, the element of rare allure,
Whispering secrets of healing and pure.

    Within the realm of medicine it dwells,
A potent ally, where science excels.
Anesthesia's embrace, it gently bestows,
Soothing the pain, calming woes.

    In art's canvas, it weaves a vibrant spell,
A prism of colors, stories it tells.
A neon oasis, a dazzling display,
Guiding the artist in a mystical way.

    But Xenon's power goes beyond the veil,
In imaging techniques, it does prevail.

A glimpse into bodies, where shadows hide,
Revealing the truth, with nothing to hide.

    A solace it brings to the tormented soul,
A balm for the wounded, making them whole.
A whisper of peace, a tranquil embrace,
Xenon's essence, a healer's grace.

    Oh Xenon, radiant and serene,
A muse for the creative, a tranquil dream.
In memories preserved, you eternally reside,
A source of enlightenment, forever by our side.

    So let us cherish this rarest of treasures,
With every breath, as life's greatest pleasures.
For in Xenon's embrace, we find solace and light,
A testament to its power, shining so bright.

# ABOUT THE AUTHOR

Walter the Educator is one of the pseudonyms for Walter Anderson. Formally educated in Chemistry, Business, and Education, he is an educator, an author, a diverse entrepreneur, and he is the son of a disabled war veteran. "Walter the Educator" shares his time between educating and creating. He holds interests and owns several creative projects that entertain, enlighten, enhance, and educate, hoping to inspire and motivate you.

Follow, find new works, and stay up to date
with Walter the Educator™
at WaltertheEducator.com

www.ingramcontent.com/pod-product-compliance
Lightning Source LLC
LaVergne TN
LVHW010603070526
838199LV00063BA/5061